Class 60s

MARK V. PIKE

BRITAIN'S RAILWAYS SERIES, VOLUME 17

Front cover image: Caught in some superb low early winter light, DB red-liveried 60019 *Port of Grimsby & Immingham* is seen approaching Didcot North Junction with 7C54, the 13.06 Oxford Banbury Road to Whatley Quarry empty stone train. 23 November 2012.

Title page image: This could almost be a brand new loco! Absolutely spotless 60015 (originally *Bow Fell*) is seen approaching Shawford with 6Z60, the 10.00 Eastleigh to Wembley Yard barrier wagon move. Strangely, considering the loco is numbered as the 15th built, it was actually the last to enter service in March 1993, despite being constructed a couple of years earlier. Back then the acceptance process was much more drawn out than it is today and many problems were encountered with the locos when new. 29 May 2012.

Page 3 image: Loadhaul 60008 passes through Newport with a westbound steel trail bound for Llanwern Steelworks. 1 July 2003.

Published by Key Books
An imprint of Key Publishing Ltd
PO Box 100
Stamford
Lincs PE19 1XQ

www.keypublishing.com

The right of Mark V. Pike to be identified as
the author of this book has been asserted in
accordance with the Copyright, Designs and Patents
Act 1988 Sections 77 and 78.

Copyright © Mark V. Pike, 2021

ISBN 978 1 80282 024 9

Typeset by SJmagic DESIGN SERVICES, India.

Introduction

The Class 60s were the last British-built diesel locomotive design and followed soon after the introduction of the America-built Class 59s. Although the private operators of Foster Yeoman, ARC and National Power were happy to order locomotives from abroad, British Rail's Railfreight sector was far more reluctant, fearing that it may encounter problems with the trades unions.

Beginning in 1989, a total of 100 Class 60 locomotives were gradually delivered to the Railfreight sub-sectors of Metals, Coal, Construction and Petroleum. This introduction was not at all straightforward though; back in those days the acceptance of rolling stock by BR was often a long-winded affair and the Class 60s were no exception, some taking almost three years to enter service owing to various teething troubles.

Once these were ironed out, the class became reasonably reliable and examples of other classes that were getting very tired at the time were progressively withdrawn. Under privatisation, all of the class were sold to English, Welsh & Scottish Railways (EWS). 2004 brought about the first Class 60 withdrawal by EWS with 60098 being stopped at Toton. Having completed only 11 years in service, it has now been stored almost twice as long as it was actually hauling trains! Towards the end of the 2000s, almost all of the locos were put into store with a seemingly very bleak future.

Then in 2010/11 came an announcement from DB Schenker, which had previously acquired EWS, that a small fleet of locos would receive a thorough 'Super 60' rebuilding to haul the company's heaviest trains, which should see them continue in service for a further 15–20 years or so. Eventually, 21 locos were refurbished for DB Schenker, another ten for Colas Rail (all since taken on by GB Railfreight) and a further four for Devon & Cornwall Railways (DCR). Despite the odd failure now and again, in their ten years of operation, the overhauled locos have proved very reliable and regularly haul some of the heaviest trains on the UK network.

Class 60s

My first photograph of one of the new Class 60s, this is 60011 *Cader Idris* at Eastleigh moving around the stabling sidings by the station. Back then, this was being used as the training loco for the Eastleigh area. The loco is still in service today and working for DB Cargo in red livery. 9 October 1990.

60009 *Carnedd Dafydd* is seen here a couple of weeks later stabled in Didcot Yard and assigned to similar crew training purposes. This loco has now been stored since the end of 2010 without a great deal of hope for it to ever run again. 27 October 1990.

60079 *Foinaven* in the appropriate Railfreight Coal sub-sector livery passes through Oxford while heading south with a merry-go-round coal train for Didcot Power Station. These locos took over from the Class 56s and 58s on these duties. 20 January 1993.

Now in DB Schenker red livery, 60079 passes through Didcot Parkway station with 6B33, the 11.25 Theale to Robeston oil train. Curiously, although this loco was put through the 'Super 60' refurbishment programme in the early 2010s, it was stored again in 2015 and has not worked since. 29 March 2012.

Still in Railfreight Petroleum livery, 60054 *Charles Babbage* rumbles through Brockenhurst in the heart of the New Forest National Park with 7O48, the 09.22 Whatley Quarry to Hamworthy Quay stone train. This loco was similarly put through the 'Super 60' refurbishment programme and is currently still in active service with DB Cargo. 18 August 2004.

This is the unusual combination of 60040 *Brecon Beacons* and 33114 *Ashford 150* at Salisbury with the DC Tours 'Mule & Otter Axeman', which was 1Z33, the 09.35 London Waterloo to London Waterloo via Okehampton, which used many different locos! These two locos were removed here at Salisbury in favour of 20169 and 20092. This was one of a series of interesting charters that ran from London to various south westerly destinations during this period, mostly on Sundays. 14 November 1992.

Class 60s on normal passenger trains are very rare; this, however, is 60033 *Anthony Ashley Cooper* in platform three at Salisbury. Having arrived here earlier in the morning, it was removed from 1V09, the 08.55 London Waterloo to Paignton; it is now seen waiting to leave with the 12.15 service to London Waterloo. This was the last day of regular Class 50 operations on this route, with 50007/033/050 all involved. However, 50033 failed the day before, so whether it was fixed or not, control at the time diagrammed 60033 on the former train instead. As you can imagine, enthusiasts at the time were very disappointed. How times change though, almost 30 years on since this day some of those same enthusiasts would probably now give a right arm to see a Class 60 on a passenger train at Salisbury! 24 May 1992.

A few years later and the same loco as in the previous image, 60033 *Tees Steel Express* has received a special British Steel powder blue livery and later still it was painted silver. It is seen in the former livery arriving at the once popular, but now redeveloped, Godfrey Road stabling point at Newport. The loco was put into store 15 years ago in April 2006 and looks very unlikely to work again. 12 October 1998.

60006 *Scunthorpe Ironmaster* was the other Class 60 painted in this attractive light blue livery. It is seen in the loop at Dawlish Warren with an Exeter Riverside to St Blazey freight. Originally named *Great Gable* when new, this loco now unfortunately has the dubious honour of being the first of the class to be cut up and disposed of, becoming nothing more than a memory during January 2020 at Toton Depot. 1 May 2000.

60049 (originally *Scafell*) is seen crawling up the incline at Dilton Marsh Halt, soon after leaving Westbury Yard with the Mondays to Fridays 6O41 10.14 Westbury Yard to Eastleigh Yard engineers' train, which, since the date of this image, has now transferred to GB Railfreight operation. In the past few years, this machine has had a very patchy history, being taken in and out of storage (mainly to be used at various locations as a yard shunter), but it has now been out of service since 2016. 7 November 2014.

A very unusual turn for 60071 *Ribblehead Viaduct*, which is seen approaching Eastleigh with 6D82, the 14.05 Eastleigh Yard to Eastleigh Works transfer move, hauling ex Southern Railway Schools Class No. 925 *Cheltenham* and a few ferry vans. The steam loco was heading to Eastleigh Works for repairs and maintenance. The Class 60 itself started life as *Dorothy Garrod* allocated to the Railfreight Coal sub-sector and has now been in storage since early 2016. 7 October 2010.

Carrying the appropriate Railfreight Construction sub-sector livery, 60019 *Wild Boar Fell* is also seen at Eastleigh but this time with a westbound stone train from Merehead Quarry for Fareham or Botley, a far more usual load than the one in the last image. Since the date of this photo, for a short period the loco has also carried the name *Pathfinder Tours 30 Years of Railtouring 1973-2003*, which was previously carried by Class 56 56038. 29 May 1992.

Having undergone refurbishment in the early 2010s, 60019, now carrying the name *Port of Grimsby & Immingham* (transferred from 60097), is seen here passing patches of Old Man's Beard on the bushes through the delightful Avon Valley on the approach to the Dundas Aqueduct, from which this image was taken, between Bradford on Avon and Bath with a diverted 6B33, the 1117 Theale to Robeston tank train. 14 February 2015.

Originally named *An Teallach*, 60091 departs the point of origin with a well-loaded 6B93, the 09.38 Eastleigh Yard to Fawley refinery. This train has since been consigned to history but thankfully the loco is still doing sterling work for DB Cargo today, now having received the name *Barry Needham*. 29 November 2012.

With the gloom of the twin tunnels in the background, 60080 *Kinder Scout* is approaching Newport station with an eastbound timber train. This loco later received the EWS maroon and gold livery and a couple of different safety-related names, but unfortunately it was stored in 2006 and has not worked since. 25 July 2001.

60070 *John Loudon McAdam*, 60026 (originally *William Caxton*) and 60028 (originally *John Flamsteed*) are seen stabled at Didcot Parkway. Differing fortunes befell these three locos later in their careers with 60070 being the second of the class to be put into storage way back in August 2004, it now being nothing more than a rusted hulk and very unlikely to work again. The other two have thankfully since been refurbished, with 60026 initially working for Colas Rail but currently part of the fleet operating with GBRf, and 60028 is also refurbished but now part of the small DCR fleet. 25 July 2001.

A thoroughly wet day sees railtour duty for Loadhaul-liveried 60007 (originally *Robert Adam*), which is seen approaching Poole past the since demolished signal box with 1Z36, the 04.46 Crewe to Poole 'Dorset Mariner' via the Hamworthy and Furzebrook branches. This is another loco happily still in service, being the first to receive the 'Super 60' refurbishment. It now operates with DB Cargo in its bright red livery and is named *The Spirit of Tom Kendell*. 15 April 2000.

Approaching an adverse signal, 60025 (originally *Joseph Lister*) is creeping slowly through Didcot Parkway with 6V53, the Lindsey to Langley Total oil train. This train has long since ceased running, as has the oil terminal at Langley, and indeed the loco. 11 April 1996.

Now renamed and painted in revised EWS livery, 60025 *Caledonian Paper* is seen passing the semaphore signals at Marchwood, on the Totton to Fawley branch in Hampshire, as it departs with 6V38, the 09.27 MoD Marchwood to Didcot stores service. The loco was stored back in October 2008 and will probably never work again. 4 August 2003.

The wheel flanges are squealing on the tight curve as 60027 (originally *Joseph Banks*) in the first version of EW&S livery passes Salisbury with 7V16, the 11.47 Fareham to Whatley empty stone train. This service is still a regular runner today, but now operates under the Freightliner banner. Unfortunately, the loco is not operating, having been stored in February 2009, just a couple of weeks after this shot was taken. 30 January 2009.

Soon after departing from Westbury, 60057 *Adam Smith* is almost travelling at walking pace up the steep 1 in 70 incline at Upton Scudamore, between Westbury and Warminster, with 7O40, the 13.35 Merehead Quarry to Eastleigh stone train. The loco was put into storage three years later in May 2006. 27 March 2003.

With its unusual white-backed nameplate, 60003 *Freight Transport Association* stands at Didcot Parkway during a run-round move. At this time, Class 60s were being used on the regular Avonmouth to Didcot Power Station MGR trains. This loco started life working for the Petroleum sub-sector and carried the name *Christopher Wren*; unfortunately it has been out of service since November 2006. 16 October 2001.

60051 *Mary Somerville* in the correct Petroleum sub-sector livery is waiting for the road at Eastleigh station with 6V13, the 12.18 Furzebrook to Hallen Marsh liquid petroleum gas tanks. This was a very popular working at the time, with a Class 60 being the normal power in the latter days of its operation. This loco was taken out of traffic in April 2010. 2 March 1994.

60024 (originally *Elizabeth Fry*) approaches Swindon with 6B33, the 13.25 Theale to Robeston tanks. Initially stored, this is one of the lucky ones. It was put through the 'Super 60' refurbishment programme in 2013 and bestowed with the name *Clitheroe Castle* (also previously carried by 60029 for a while) but was stored again between February 2016 and April 2019 for rectification work. It is now currently hard at work with DB Cargo. 23 April 2009.

This is 60063 *James Murray* passing through Newport station with one of the well-known Llanwern to Port Talbot iron ore trains that formerly utilised three Class 37 locos and after that two Class 56s. They have since ceased running altogether. However, the now-nameless loco is still regularly seen here today, working for DB Cargo in its bright red livery. 12 April 1995.

18 years after the last photo was taken, 60063 is now fully refurbished and approaching Hawkeridge Junction, near Westbury, with 6A83, the 14.17 Avonmouth Bennet's Siding to West Drayton stone train. This was quite an unusual working for the class at the time. 11 January 2013.

60081 *Isambard Kingdom Brunel* is seen in ex-works GWR green livery at Hamworthy Goods, near Poole, running round its steel train. This was the site of the original Poole station in use from 1847 until 1896 when the new Poole station, at the location we know today, was opened. It must surely hold some sort of record as being one of the few lines still in use today (albeit rarely in the 21st century) that was closed to passengers back in the 19th century! 11 August 2000.

A short while after the previous image, 60081 *Isambard Kingdom Brunel* is just getting underway and slowly approaching Lake Level Crossing with 6V99, the 11.35 Hamworthy Quay to Cardiff Tidal. Although appearing like double track, the line to the right of picture is just a long siding with a buffer stop behind the camera. This was one of the loco's first workings since it was raised from obscurity as just another Class 60 and unveiled to the public at the Old Oak Common open day a few days before. Originally named *Bleaklow Hill,* the loco suffered a catastrophic engine failure in 2005 and was stored as a result. It is one of only a few of the class officially withdrawn from service (most others are just classed as stored), but recent developments have seen the loco acquired by Locomotive Services with the intention of restoring it cosmetically for display in a new museum at Margate, with an option of returning it to working order at some point in the future. 11 August 2000.

Just under one month old, 60067 *James Clerk Maxwell* is getting underway from a signal check at Reading with 6E50, the 09.45 Saturdays-only Langley to Lindsey oil refinery empty tanks. Just 15 years later, in April 2006, the loco was put into storage and has not worked since, which incredibly makes it the same amount of time in storage as in front line service! 19 October 1991.

Another example that has had a short working life, Railfreight Petroleum-liveried 60064 *Back Tor* passes through Reading with the 6V53 Lindsey to Langley oil train. Again having completed just 15 years in service, it was stored in January 2006. 21 July 1992.

At the other end of the scale, 60029 *Ben Nevis* with the Transrail 'big T' decals passes through Cardiff Central with a Llanwern to Port Talbot empty iron ore train and is now enjoying a second lease of life. Although this loco was in storage for a number of years, it was one of the few selected by DCR in 2014 to undergo the 'Super 60' refurbishment and, as a result of this, it should be in service for at least the next ten years or so. 5 June 1996.

Looking more like a brand new locomotive, 60029 *Ben Nevis* is seen again 24 years on from the date of the last image, passing Salisbury with 6Z90, the 07.09 Southampton Up Yard to Westbury stone empties. In a nice move by DCR, the company has restored the original name applied when the loco was new back in 1991. 9 October 2020.

An unmistakeable location for 60043 (originally *Yes Tor*) as it powers beneath the fine semaphore signals that stood on the approach to Barnetby with an eastbound iron ore train. This area was completely resignalled during the late 2010s, with the semaphores and signal box now just history. Unfortunately, the loco is now almost history itself as it was stored back in December 2008 with no sign of movement since. 7 March 2003.

In a very weather worn version of the dark blue Mainline Freight livery, this is 60011 (originally *Cader Idris*) approaching Langley station on the Great Western Main Line with 6M20, the 10.20 Whatley to St Pancras stone train. This view, like most others on the Great Western, has now changed totally with the advent of electrification. 26 February 2009.

Just two years later and 60011 became the first of the class to carry DB Schenker red livery. Soon after its repaint, the spotless machine is seen here approaching Cam & Dursley, just south of Gloucester, with 6B13, the 05.07 Robeston to Westerleigh tanks. Curiously, the loco was not part of the 'Super 60' programme but it is still in everyday service with DB, though now un-named. 13 January 2011.

Carrying Mainline decals, but still in two-tone grey, 60018 *Moel Siabod* passes Kensington Olympia with a southbound stone train formed of a uniform rake of four-wheeled Yeoman PGA hoppers (now withdrawn) from the distribution yard at Acton. 4 June 1996.

Another eight years on and an un-named 60018, now in the revised version of EWS livery, is still hauling stone trains. It is seen passing Millbrook with 7O48, the 09.23 Whatley to Hamworthy. This loco was stored in May 2009 and has since become one of the few officially withdrawn from service, as opposed to simply stored. 16 January 2004.

Possibly the loco with the shortest working life, 60098 *Charles Francis Brush*, in Railfreight Construction livery and having almost completed six months of service, is seen passing a damp Leicester station with a southbound stone train. Despite this being the 1,000th loco built by Brush at Loughborough, it was the first of the class put into store in April 2004, after completing just over 11 years in service. It has now been stored for six years longer than it was in service and, during the intervening 17 years or so, it has been progressively robbed for spares to help keep other locos running. 16 June 1993.

With the EWS big 'beastie' sticker on the bodyside, 60060 *James Watt* heads west through the old Reading station (which has since been completely rebuilt) with 6V18, the 12.43 Hither Green to Whatley Quarry. The loco was put into storage during February 2009, just two months after this image was taken. 5 December 2008.

60017 (originally *Arenig Fawr*) forms up 6V41, the 14.47 infrastructure service to Westbury next to Eastleigh station. Owing to the many problems encountered when commissioning these locos when new, this turned out to be the first of the class accepted into service at the end of October 1990. 5 July 2016.

Immaculate in Colas Rail livery, 60085 (originally *Axe Edge*) pauses at Reading with 6V62, the 10.56 Tilbury Riverside to Llanwern steel empties. Prior to transferring over to Colas, this loco also carried the name *Mini - Pride of Oxford*. It has since been taken on by GB Railfreight. 14 January 2015.

Exiting the tunnel, 60083 *Mountsorrel* is approaching Southampton Central with 6W53, the 08.45 Eastleigh Yard to Furzebrook gas train. Originally named *Shining Tor*, this loco was stored in November 2008. 4 December 2003.

Ex-works 60054 (originally *Charles Babbage*) is captured passing Hinksey, just south of Oxford, with 6E55, the 13.35 Theale to Lindsey oil refinery. Currently not named, the loco is still hard at work for DB Cargo. 4 November 2011.

In revised EWS livery, 60035 (originally *Florence Nightingale*) is approaching Eastleigh with 6M52, the 1132 Southampton Eastern Docks to Castle Bromwich (Jaguar). At the time, this was a very rare working for the class and was probably a result of the non-availability of the usual Class 66. This machine has had a chequered history in recent years, going through a few phases of storage and active service as and when required, usually for use as a yard shunter. However, it has now been stored since January 2016 so DB may have finally given up with it! 18 November 2015.

An imposing view of 60031 *ABP Connect* awaiting a driver at Westbury with an unidentified loaded stone train. For a while, this loco was recognisable by the significant dent in the bodywork above the bufferbeam. Originally *Ben Lui*, it was stored back in March 2009 and is not expected to operate again. 8 October 2003.

This is 60065 *Spirit of Jaguar* approaching Mount Pleasant level crossing with 6B43, the 09.39 Eastleigh Yard to Southampton Western Docks. At the time, this was the yard shunting loco at Eastleigh and this short out and back trip working was acting as a bit of a training run, as can be judged by the amount of people in the cab! The derelict looking yard above the loco (formerly Bevois Park) has since been reopened to stone trains. Originally named *Kinder Low*, the loco itself is one of just a couple of un-refurbished examples still active today, albeit for yard shunting and occasional driver training runs. 21 August 2013.

Another example that received the so called big 'beastie' EWS logo on the bodysides, 60073 *Cairn Gorm* also has the addition of an enlarged '073' on the front end. It is seen passing Tilehurst East Junction with 6B33, the 11.25 Theale to Robeston tank train. The loco was put in to storage in March 2011. 10 September 2009.

Looking smart in Colas Rail livery, 60002 (originally *Capability Brown*) is just west of Patchway station with the short lived 6M18, the 06.58 Margam to Washwood Heath steel train. This loco also carried the name *High Peak* during its days with EWS but is now in GBRf livery and recently named *Graham Farish 50th Anniversary 1970-2020*. 27 July 2015.

Still carrying the obsolete Railfreight Petroleum decals, 60014 *Alexander Fleming* gets underway at Bletchley station on the West Coast Main Line with 6H55, the 09.37 Bletchley to Peak Forest empty stone train. One of the few officially withdrawn examples, it was put into storage in December 2008. 20 February 2003.

In full but rather faded Railfreight Coal livery, 60090 *Quinag* heads south through the single island platform at Micheldever station on the South Western Main Line with an empty sand hopper train bound for Wool, Dorset. The loco was stored in April 2008 at Toton and remains there to this day with no signs of further use on the horizon. 15 April 2003.

The pioneer of the class, 60001 (originally *Steadfast*) arrives at Westbury with 7V82, the 05.02 Crawley to Merehead empty stone train. Although the loco was stored for some time in the late 2000s, it was one of the batch chosen to undergo the 'Super 60' overhaul and it is still very much at work with DB Cargo today. It makes an interesting comparison with 59205 stabled behind it. 7 November 2014.

Transrail-branded 60032 *William Booth* heads south with a steel train at Shipton by Beningborough on the East Coast Main Line, a few miles north of York. Some readers will doubtless be aware that this image was taken from The Sidings bed & breakfast hotel formed of former railway carriages situated right next to the line. Although I have not visited for some years now, I think it is still a great place to stay if you are a railway enthusiast! The loco was one of just a few that received an official naming ceremony when new, which was held at Nottingham station on 23 November 1990 by the grandson of William Booth, the loco then entering service a week or so later for the Railfreight Coal sub-sector. It has now been in store at Toton since July 2006 and not likely to work again. 5 April 2002.

Displaying no sub-sector branding at all, this is 60068 *Charles Darwin* approaching Millbrook with 7O48, the 09.22 Whatley Quarry to Hamworthy stone train. Stored in September 2009, it is one of many of the class now gradually rusting away in the yard at Toton. 5 March 2004.

Always an elusive one for me, 60058 (originally *John Howard*) is passing through Newport station with a steel train bound for Llanwern. Also of interest is former Southern Region Class 09 09015 waiting for a signal to proceed with one of the short trip workings that operated at the time between East Usk Sidings and Alexandra Dock Junction. No doubt the driver was thankful of the extra turn of speed from his Class 09 compared to a 08! The Class 60 was stored during January 2008 at Toton and has not worked since. 13 October 2005.

A selection of images now illustrating the one-off livery that adorned 60074 (originally *Braeriach*). This first one shows the loco prior to becoming a celebrity as it passes through Eastleigh with 6O41, the 10.14 Westbury Yard to Eastleigh East Yard engineers' train. 23 December 2003.

Seven years later, and with a new name and an attractive powder blue livery, 60074 *Teenage Spirit* has just passed through West Drayton with 6M20, the 10.20 Whatley to St Pancras stone train. 30 September 2010.

60074 *Teenage Spirit* is seen again, this time heading through Clapham Junction (Windsor Lines) with a westbound empty stone train bound for one of the Somerset quarries. At this time, Class 60's were becoming much less common in the London area. 27 January 2011.

Finally, 60074 *Teenage Spirit* is passing through Cam & Dursley, just south of Gloucester, with 6B13, the 05.07 Robeston to Westerleigh tanks. In 2021, this is still a solid Class 60 working. Recently, this loco has received yet another one-off Puma Energy livery of dark grey; it also carries the name *Luke*. 28 February 2012.

A couple of before and after shots of 60092 *Reginald Munns*. It is seen awaiting the road at Millbrook with 6B44, the 12.07 Southampton Western Docks to Eastleigh East Yard trip working. The class has always been rather rare on Intermodal services. This was one of the lucky ones to undergo 'Super 60' treatment and is currently in service with DB Cargo. 30 March 2004.

Now in DB Schenker red and un-named, 60092 is basking in some lovely autumn sunshine at Swindon whilst shunting the recently arrived 6B49, the 06.46 Llanwern Sidings to Swindon Stores. 9 October 2015.

Above: In the popular black and orange Loadhaul livery, 60007 (originally *Robert Adam*) is about to leave Newport after a crew change with a westbound steel train bound for Llanwern. This was the first loco to undergo the 'Super 60' refurbishment and has also been renamed *The Spirit of Tom Kendell*. 7 December 2005.

Left: Also in Loadhaul livery, 60008 *Gypsum Queen II* is exiting the gloom of Newport Tunnel with a westbound steel train bound for Llanwern. It was originally named *Moel Fammau* from new and, despite it later receiving EWS maroon and gold livery and being renamed *Sir William McAlpine*, it was stored back in June 2007. 1 July 2003.

This is 60010 (originally *Pumlumon Plynlimon*) crossing Holes Bay causeway soon after passing Poole with 7O48, the 09.23 Whatley Quarry to Hamworthy stone train. Although not looking much like a causeway now, the land either side of the line here originally formed part of Holes Bay itself, with water lapping up to the line on what was originally a low embankment until land reclamation in the late 1960s and early 1970s. The loco has since been refurbished and now forms part of the DB Cargo fleet. 12 December 2007.

60097 *Port of Grimsby & Immingham* (the name since transferred to 60019) passes through the centre road, unusually in the southbound direction, at Kensington Olympia with a stone train from Acton. This loco was originally named *Pillar* and unfortunately became part of the now large collection of stored Class 60s at Toton Depot during August 2008. 9 July 2003.

60011 (originally *Cader Idris*) and 66182 are seen departing from Westbury Yard with the 10.14 to Eastleigh East Yard. This location is right at the start of the steep incline that continues for around two miles to the summit at Upton Scudamore, near Warminster. As steep as 1 in 70 in places, heavy stone trains very often required banking assistance in the past. In this instance, it was just hauling the Class 66 dead in train and not for any extra power. 60011 is still in service with DB, although interestingly it has not had the 'Super 60' treatment. 12 June 2012.

The pioneer again, 60001 (originally *Steadfast*) is stabled in a bay platform at Reading. It was quite common at this time to see a Class 60 here as often the loco off the overnight tanks to Langley ran back to layover until the afternoon. Although this loco was delivered to BR brand new at the end of June 1989, it was not actually accepted by BR for another two years in August 1991 because of many technical details encountered with the new locos. So, despite carrying the lowest number, it was surprisingly the 53rd loco to enter traffic! It was also named *The Railway Observer* for a short time. It has now been refurbished and is in everyday use with DB Cargo, although now nameless. 3 March 2000.

The now fully refurbished 60001 is seen passing North Staffordshire Junction, south of Derby, with 6M57, the 06.43 Lindsey oil refinery to Kingsbury Sidings. The area off to the right of the picture was once the huge Willington Power Station complex but has now long since been closed, although the disused cooling towers still stand at the time of writing. 11 February 2016.

60051 (originally *Mary Somerville*) is seen heading north through the cutting prior to entering the 1,280m-long Wickwar Tunnel, between Yate and Charfield, with 6E41, the 11.41 Westerleigh to Lindsey. Unfortunately, even though this loco looks clean and tidy, it was put into storage at Toton during the same month as this image was taken and has been there ever since. 8 April 2010.

This is 60021 (originally *Pen-y-Ghent*) crossing the famous Ribblehead Viaduct with 6J37, the 12.58 Carlisle Yard to Chirk Kronospan log train. Just two months after this image was taken, the loco was acquired by GBRf along with all the other Colas Rail examples and has now been painted in its blue and orange house colours. As a nice touch, it has regained its *Penyghent* name, this time with the nameplates fashioned in the original Class 44 'Peak' style carried by 44008. 29 May 2018.

Heading down the centre road, 60042 *The Hundred of Hoo* passes through Eastleigh with 7O48, the 09.23 Whatley Quarry to Hamworthy stone train. Originally named *Dunkery Beacon*, the loco was stored in March 2008. Also of interest are the two Class 66s in view here, the one on the left (66250) has since become 66789 and is now working for GBRf in BR large logo livery, and the one to the right (66031) is now working for Direct Rail Services in its dark blue livery. How times change! 13 March 2003.

60095 (originally *Crib Goch*) is approaching Sherrington foot crossing in the Wylye Valley, between Salisbury and Warminster, with 6V30, the 10.39 Eastleigh to Westbury long welded rail train. This was one of the locos taken on by GBRf in July 2018. It is currently nameless, although now in GBRf livery. 7 August 2015.

60059 *Swinden Dalesman* passes Tamworth High Level with 6E54, the 10.39 Kingsbury oil sidings to Humber oil refinery. This loco was originally named *Samuel Plimsoll* but is now reunited with the name it carried during its period of time spent in the orange and black Loadhaul livery. 5 February 2016.

60070 *John Loudon McAdam* is seen stabled between duties at Westbury. This was one of only three locomotives (with 60050 and 60064) to carry the Loadhaul decals over the original Railfreight triple grey livery, rather than the full black and orange treatment. After the ravages of 17 years of storage (August 2004) in the open at Toton Depot, this loco is now very unlikely to ever work again, being not much more than a shell. It was the second member of the class to be stored. 3 September 2003.

60062, originally *Samuel Johnson*, passes Eastleigh with 7O48, the 09.23 Whatley Quarry to Hamworthy Quay stone train. The loco has since undergone 'Super 60' refurbishment and is now named *Stainless Pioneer*, which was first carried by a Class 37 and then a Class 56. 60062 is now in every day service with DB. 18 February 2004.

With the famous landmark of the 123 metre (404ft) high spire of the cathedral prominent in the background, 60002 *High Peak* is seen soon after passing Salisbury with 7V16, the 11.47 Fareham to Whatley empty stone train. Towards the rear of the train to the right used to be the old Salisbury steam shed, closed when steam ended in 1967. Originally named *Capability Brown*, the loco has now been refurbished, firstly working for Colas Rail and then from July 2018 for GBRf. It now carries the name *Graham Farish 50th Anniversary 1970-2020*. 18 January 2006.

Ex-works and fully refurbished 60019 *Port of Grimsby & Immingham* is captured approaching Didcot North Junction with 7C54, the 13.06 Oxford Banbury Road to Whatley Quarry empty stone train. Originally named *Wild Boar Fell*, it also carried the name *Pathfinder Tours 30 Years of Railtouring 1973-2003* for a short while. 23 November 2012.

Having just come off the freight only line (as it was back then) from Romsey, via Chandlers Ford, Loadhaul liveried 60007 (originally *Robert Adam*) heads south through Eastleigh station with 7O48, the 09.23 Whatley Quarry to Hamworthy Quay stone train. This was later the first loco to be put through the 'Super 60' programme. 7 February 2004.

With one of its first duties since being refurbished and renamed, 60007 *The Spirit of Tom Kendall* is seen in sparkling ex-works condition passing through Purley on Thames, on the Great Western Main Line, with 6B33, the 13.25 Theale to Robeston tanks. This view is now impossible with the coming of the electrification that arrived in this area during 2016/17. 22 September 2011.

The Spirit of
Tom Kendell

A little further along the Great Western Main Line than the previous image at South Moreton, near Didcot, is EWS stickered 60013 *Robert Boyle*, once again with the same train, the 6B33 13.25 Theale to Robeston tanks, which is unusually taking the westbound fast line. This slow moving train is more commonly routed along the westbound local line seen behind the loco. 60013 was unfortunately sent to storage at Toton just seven months after the date of this image. 7 April 2011.

Despite looking a very rural location, this is 60045 *The Permanent Way Institution* passing the Lower Test Marshes and approaching Redbridge, just south of the urban sprawl of Southampton, with 6O41, the 10.14 Westbury to Eastleigh East Yard. This service has since transferred to GBRf operation. The loco was originally named *Josephine Butler* and was stored in December 2014. 4 September 2013.

On a beautiful spring morning, 60059 *Swinden Dalesman* is seen passing the Freightliner depot at Southampton Maritime with 6B94, the 12.00 Fawley to Eastleigh East Yard. Originally named *Samuel Plimsoll*, the now fully refurbished loco still continues to work for DB. 25 April 2014.

The driver of 37190 on an eastbound steel train is on the phone to the signaller at Newport station as 60020 *Great Whernside* passes by with a Llanwern to Port Talbot empty iron ore train. The Class 37 has since entered preservation and the Class 60 has now been refurbished and renamed *The Willows*. 16 April 1992.

With an excellent uniform train of oil tanks, this is refurbished 60020 *The Willows* approaching North Staffs Junction with 6E54, the 10.39 Kingsbury Sidings to Humber oil refinery. The small Willington station can just be made out at the rear of the train. 11 February 2016.

In complete contrast to the previous image, freight trains do not come much smaller than this! Coming off the Reading line at Basingstoke is 60035 (originally *Florence Nightingale*) with just one OBA open wagon as 6O26, the 10.19 Hinksey Sidings (Oxford) to Eastleigh East Yard. The train still runs but is now in the hands of Colas Rail. The loco was stored in January 2016, somewhat later than many after being regularly used as a yard shunting loco at various locations. 5 October 2013.

Approaching Redbridge and the junction with the main Waterloo to Bournemouth/Weymouth line, this is 60049 (originally *Scafell*) with the well-loaded 6O41, the 10.14 Westbury to Eastleigh. This is another one of the class that has been in and out of storage quite a few times, as it was used for shunting various DB Schenker yards around the UK. In fact, it was being used as the Westbury shunter on this day but was pressed into service because of a lack of Class 66s. It has now been stored since February 2016, so it looks doubtful it will work again. 1 April 2011.

Cardiff Central used to be a good place to observe the many steel trains that passed through. This is 60047 (originally *Robert Owen*) with an eastbound train bound for Llanwern. This is one of the ten locos refurbished for use by Colas Rail in 2014, but it has since been taken on by GBRf and is currently un-named. 7 September 2006.

The now refurbished 60047 is seen during its time working for Colas Rail as it passes by the well-known location of Dawlish with 6C22, the 13.02 Par to Westbury engineers' train. Class 60s were once fairly regular visitors to Devon and Cornwall but their appearances these days usually cause quite a stir. The sea wall has since been rebuilt along this section to try and stop the damage caused by high seas. 30 January 2016.

Looking quite smart in the first version of EW&S livery, this is 60004 (originally *Lochnagar*) running light engine through Newport station. This loco was stored at Toton Depot in November 2009 but has since been officially withdrawn, with probably just the prospect of the cutting torch to come. 13 June 2003.

A very work-weary 60044 (originally *Ailsa Craig*) is seen stabled at Didcot Parkway awaiting its next duty. This is one of just three examples (with 60011 and 60078) to receive the dark blue Mainline Freight livery, sometimes referred to as aircraft blue. Upon the EWS takeover in the mid-1990s, rather than do a full repaint the company often just branded locos with large stickers on the bodyside such as this one. Interestingly, it still carried its original name when in full Mainline blue, but you can see the patch left on the bodyside where it had been removed quite recently. 6 December 2007.

12 years later, and now looking much more respectable than in the previous image after the 'Super 60' treatment, 60044 *Dowlow* negotiates its way through Eastleigh station with 7V16, the 11.47 Fareham to Westbury empty stone train. This picture was taken just a couple of months before Freightliner took over the operation of this service. 17 September 2019.

Back in the days when the class was used extensively on the Avonmouth to Didcot Power Station MGR coal trains, this is Mainline stickered 60094 *Tryfan* arriving at Didcot Parkway with yet another trainload. Usually, these trains stopped short of the station and the loco would come in to the station to run round, but occasionally the whole train would come into the station as seen here. 11 April 1996.

Ten years later, 60094, now named *Rugby Flyer*, is seen passing Reading West with 6O26, the 10.19 Hinksey Sidings to Eastleigh East Yard infrastructure service. This train has since been operated by Colas Rail. The loco was declared surplus by EWS and stored in February 2010. 14 September 2006.

Also at Reading West, but looking in the opposite direction from the previous image, this is 60024 (originally *Elizabeth Fry*) with 6B33, the 13.25 Theale to Robeston tanks. This view is literally now impossible, not only because of the overhead electric wires, but also for the small fact that the footbridge from which it was taken no longer exists! Since being refurbished and at work for DB, it now carries the name *Clitheroe Castle*, which was once carried by 60029. 16 April 2009.

Passing the lovely setting of Greenland Mill, with its fine weir in the foreground, this is 60091 *Barry Needham* commencing its run through the Avon Valley on the approach to Bradford-on-Avon with the diverted 6B33, the 11.11 Theale to Margam. Originally named *An Teallach*, the loco now carries the name once carried by 56115. 18 April 2015.

Presumably chosen as it was the latest loco to be refurbished, this is 60015 passing Earls Court (since demolished) with the 6Z30 Dollands Moor to York. In the middle of the set of barrier wagons is Class 395 Javelin 395019, which was being conveyed to the National Railway Museum for display at the Railfest event being held that weekend. 30 May 2012.

60076 (originally *Suilven)* passes the canal lock gates at the classic location of Crofton on the Berks and Hants line with the diverted 6V62, the 10.44 Tilbury Riverside to Llanwern steel train. It currently carries the unofficial name *Dunbar* and is now in operation with GBRf. 8 April 2015.

Soon after exiting the 2.5 mile long Chipping Sodbury Tunnel, between Bristol Parkway and Royal Wootton Bassett, 60013 *Robert Boyle* is in charge of 6B33, the 13.25 Theale to Robeston tanks. 60013 was put into store in May 2009 and has not worked again. 5 September 2009.

A gleaming 60007 *The Spirit of Tom Kendall* is seen passing over the River Usk at Newport with 6B13, the 05.00 Robeston to Westerleigh oil train. Overhead wiring has now ruined this shot. 29 March 2012.

Coming past the Freightliner terminal at Millbrook is 60093 *Jack Stirk* with 6O41, the 10.14 Westbury to Eastleigh East Yard. This has since become yet another example in the Toton Depot storage ranks, being consigned there in January 2009, and now gently rusting away. 16 June 2003.

Having come off the Fawley branch, just beyond the footbridge in the distance, this is 60011 (originally *Cader Idris*) approaching Totton station with the 6Y32 08.24 Fawley to Holybourne tanks, which was an unusual duty for one of the class. This train has since stopped running, but fortunately the loco continues to work for DB Cargo. 15 June 2012.

Over the years, the class has often been used to haul various railtours across the country. Having just taken over from 37406 and 37401, this is 60091 *An Teallach* still in full Railfreight Coal livery and waiting to depart from Salisbury with 1Z37, the 05.04 Derby to Weymouth 'The Solent Syphons' charter, which it worked from here to Southampton up goods loop. From there, it continued on the rear of the train to Weymouth hauled by Class 57/0 57002. 9 April 2005.

At a location that normally only sees electric units, this is rather an unusual selection of locos at the south coast terminus of Weymouth! 60091 *An Teallach* along with 57002, 37406 and 37401 were all at Weymouth in conjunction with the charter seen in the previous image. The Class 60 and 57 later returned on a light engine move to Eastleigh Yard. 9 April 2005.

Another railtour, this time seeing 60021 (originally *Pen-y-Ghent*) approaching Cardiff Central with 1Z42, the 13.50 Margam to Crewe 'Valley Vostock' charter that visited various lines in the Welsh Valleys. As well as the use of the Class 60, this tour also featured 47197, 56038, 57007 and 66614 at various stages. 60021 is still in service, having been refurbished in 2014, firstly for Colas Rail but now operating for GBRf. 7 February 2004.

Just over ten years later, during its time with Colas, 60021 is seen at the rebuilt Reading station with 6V62, the 11.22 Tilbury to Llanwern steel train. 1 October 2014.

This is 60008 *Gypsum Queen II* approaching Holton Heath, just east of Wareham, with a short sand hopper train bound for Wool. This traffic eventually went over to Freightliner haulage but has now ceased running altogether. This loco started life as *Moel Fammau* and eventually *Sir Robert McAlpine* but has now been in storage since June 2007. To the left of picture, almost the whole area was originally part of the Royal Navy cordite factory that was set up during World War One to provide propellant for the navy's various armouries, and this was continued during World War Two. Dealing with notoriously volatile substances such as nitro-glycerine, there were inevitably a few serious accidents there during its time. After the war it was closed and, after decades of ensuring the site was safe, most of the area has become a pleasant nature reserve and a not quite so pleasant industrial estate. 27 January 2003.

Just a couple of months prior to Freightliner taking over the Mendip stone contract in November 2019, this is 60024 *Clitheroe Castle* thumping its way through Radley station, just south of Oxford with 6C58, the 11.45 Oxford Banbury Road to Whatley Quarry empty stone train. This loco started life as *Elizabeth Fry*. 10 September 2019.

It is very unlikely that any train on the UK network would require the power of more than one Class 60 and most instances usually occur because of the requirement of needing a loco elsewhere, or as a result of loco failure. The next four images show a few of these occurrences. This one is 60066 (originally *John Logie Baird*) and 60083 *Mountsorrel* passing the Freightliner terminal at Millbrook with the 6B94 Fawley to Eastleigh tanks. These two locos have had very different fortunes over the last few years; 60066 was the last DB loco to undergo the 'Super 60' refurbishment, but unfortunately 60083 (originally *Shining Tor*) was stored back in November 2008 and has not worked since. 2 December 2003.

Well, what can I say, this was quite incredible! A potential 9,300hp in the form of 60049 (originally *Scafell*), 60035 (originally *Florence Nightingale*) and 60071 (originally *Dorothy Garrod*) are seen passing the same spot as the previous image with 6O41, the 10.14 Westbury to Eastleigh engineers' train. In reality, only the leading loco was working, but to date this is the only time I have ever seen three Class 60s on the front of a train and I doubt it will happen again. Unfortunately, all these locos have since been stored; 60049 in February 2016 and 60035 and 60071 both in January 2016. 20 April 2004.

Another surprise and amazingly on the same day as the last image. This is 60032 *William Booth* and 60040 (originally *Brecon Beacons*) passing Millbrook in the opposite direction a few hours later with the return 6V41 14.47 Eastleigh East Yard to Westbury. It is quite obvious from the size and weight of the train in tow that 6,200hp was not required here! 60032 was stored as long ago as July 2006, while 60040 has fared much better, being refurbished in 2011/12 and now in normal service with DB. It also now carries the name *The Territorial Army Centenary*. 20 April 2004.

60045 *The Permanent Way Institution* and 60049 (formerly *Scafell*) are approaching Freemantle footbridge on the approach to Southampton Central with 6O41, the 10.14 Westbury to Eastleigh. This combination was used to get an extra loco to Eastleigh and not because of failure. Note the differing original EW&S and revised EWS liveries. 60045 (originally *Josephine Butler*) was stored in December 2014 and 60049 was stored in February 2016. 5 September 2013.

Rounding the tight curve from Eastleigh East Junction, refurbished 60007 *The Spirit of Tom Kendall* is taking the Romsey line via Chandlers Ford with 7V16, the 11.47 Fareham to Westbury. Up until the early 2000s, this short line connecting Eastleigh with Romsey was just freight only after being downgraded in the late 1960s. However, more recently, there has been a regular passenger service operated by South Western Railway, which came about with the reopening of the only station on the route at Chandlers Ford. 4 May 2016.

A very work-stained 60034 (originally *Carnedd Llewelyn*) is seen at the former Godfrey Road stabling point at Newport. This always used to be a popular location with enthusiasts as there were often a good amount of locos stabled between duties. With the rebuilding of the station in the 2010s, the area was totally redeveloped to accommodate an extra platform, with any loco stabling now being at Alexandra Dock Junction, just west of the station. The loco was stored at Toton Depot in August 2008 and has not worked since. 5 June 2003.

Still sparkling after its recent 'Super 60' overhaul, 60087 *CLIC Sargent* approaches Twyford station with 6V62, the 11.22 Tilbury to Llanwern steel train. Originally named *Slioch*, it was later renamed *Barry Needham* (which is now carried by 60091). Since its transfer to GB Railfreight in 2018, it has carried the unofficial name *Bountiful*; quite a collection! 17 September 2014.

60008 *Gypsum Queen II* catches the low sun on the approach to Eastleigh with 6V38, the 11.39 Marchwood to Didcot Yard MoD train. The loco was originally named *Moel Fammau* and later renamed *Sir William McAlpine*; it was stored back in June 2007. 30 November 2004.

60039 (originally *Glastonbury Tor*) is seen approaching Winchester with 6O26, the 10.19 Hinksey Sidings to Eastleigh East Yard engineers' train. It is interesting to note that the area below the bufferbeam on this loco is painted yellow. This was one of the lucky examples chosen for the 'Super 60' treatment and is now in full time service with DB and carrying the name *Dove Holes*. 28 August 2009.

Coming around the curve through Newport station, this is 60077 (originally *Canisp*) with 6B13, the 05.00 Robeston to Westerleigh oil train. As one of the heaviest trains operated by DB, this has almost exclusively been Class 60 hauled for many years now; unfortunately not by this loco any more, as it was stored back in December 2009. 1 July 2003.

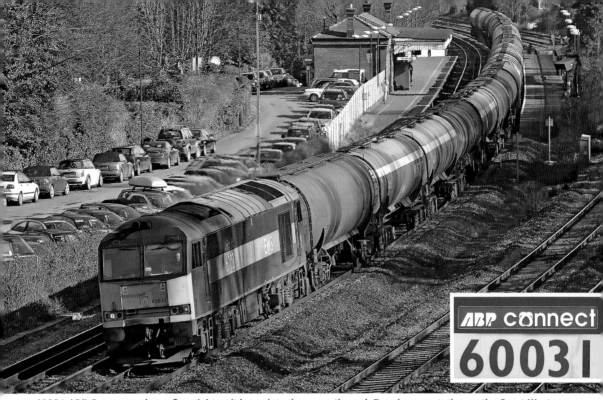

60031 *ABP Connect* makes a fine sight as it leans into the curve through Pangbourne station on the Great Western Main Line with 6B33, the 13.25 Theale to Robeston tanks. This is another impossible view nowadays with the electrification wires interrupting the view. Originally named *Ben Lui*, this loco was put into storage in March 2009, very soon after this shot was taken, and has not worked since. 4 March 2009.

In recent years, quite a few rakes of charter stock have been based at Eastleigh, but when charters run, they often end up terminating late in the evening in the London area, meaning that the stock is often returned the following day to Eastleigh. This is usually carried out by a DB Class 66 or 67 but, on this occasion, the duty unusually fell to 60017 (originally *Arenig Fawr*). During its time with EWS, the loco also carried the name *Shotton Works Centenary Year 1996* for a while. It is seen approaching Basingstoke with 5O61, the 10.00 Wembley Yard to Eastleigh empty stock. 21 May 2013.

Waiting for the road on the reversible through line at the original Reading station, complete with EWS stickers, this is 60013 *Robert Boyle*, at this time most unusually carrying a red-backed nameplate, with 6M20, the 10.20 Whatley Quarry to St Pancras. Although looking very presentable, the loco was stored just two months later. 10 March 2011.

From the well-known viewpoint of Campbell Road Bridge at Eastleigh, 60087 *Barry Needham* heads south with 7O48, the 09.23 Whatley Quarry to Hamworthy stone train. The name carried at the time by this loco originally adorned 56115 and has since been transferred to 60091. 60087 (originally *Slioch*) is now operating with GBRf and carrying the unofficial name *Bountiful*. 20 April 2005.

This is the bizarre combination of 60087 *CLIC Sargent*, 70804 and 70810 on the rear passing Wimbledon West Junction with 6Z70, the 08.50 Rugby Depot to Eastleigh Yard. This was one of 60087's first workings after refurbishment for Colas Rail. 4 July 2014.

The amazing amount of heat haze emanating from 60019 *Port of Grimsby & Immingham* shows just how hard it is working to get 6B13, the 05.00 Robeston to Westerleigh oil train, on the move again upon departure from a signal check on the approach to Severn Tunnel Junction. Originally named *Wild Boar Fell* and later *Pathfinder Tours – 30 Years of Railtouring 1973–2003*, it gained its current name upon withdrawal of 60097, the original carrier. 15 February 2016.

Complete with a yellow-painted mini snowplough, 60016 *Rail Magazine* is seen on display at the Old Oak Common open day. Originally named *Langdale Pikes* when new, it was renumbered 60500 in November 2004 to commemorate the 500th issue of the magazine of the same name. Unfortunately, it was put into storage at Toton in April 2008 and has not worked since. 5 August 2000.

Wearing the original EW&S maroon and gold livery, 60040 (originally *Brecon Beacons*) awaits its next turn of duty in the sidings at Didcot Parkway. The loco later went on to carry a one off maroon colour when it was named *The Territorial Army Centenary* in the late 2000s. 19 November 2002.

Nine years later, 60040, now named *The Territorial Army Centenary* and painted in the aforementioned attractive maroon livery, is seen passing Millbrook with a rather lengthy 6V41, the 14.47 Eastleigh East Yard to Westbury. Freightliner's 66563 is waiting to enter Millbrook container terminal in the background. 4 May 2011.

Now in its latest livery of DB red, but still carrying *The Territorial Army Centenary* nameplates, 60040 is seen passing Romsey with 6O41, the 10.14 Westbury to Eastleigh East Yard. 17 September 2012.

Approaching Southampton Central, this is 60016 *Rail Magazine* with 6V38, the 11.39 Marchwood to Didcot Yard MoD train. This was a few months prior to it receiving the number 60500, as mentioned earlier. 17 February 2004.

16 years later, at exactly the same spot as the last image, this is recently refurbished 60055 *Thomas Barnardo* with 6Z90, the 07.27 Southampton Up Yard to Eastleigh Yard via a run-round in Southampton Up Loop. This is quite a sporadic train and only a short term one, unfortunately. Although it is essentially just plain grey, the DC Rail livery does seem to quite suit the class. 10 September 2020.

60091 (originally *An Teallach*) is seen here on the approach to Basingstoke with 6V27, the 13.30 Eastleigh East Yard to Hinksey Sidings infrastructure service, which is now Colas Rail operated. 4 December 2012.

A very battered and weather worn 60049 (formerly *Scafell*) is seen during shunting operations at Westbury. Since the takeover of the Mendip operations by Freightliner in November 2019, the yards at Westbury are still actually operated by DB using its own locomotives, so it is just possible a Class 60 could put in an appearance here in future. This loco was stored in February 2016. 7 November 2014.

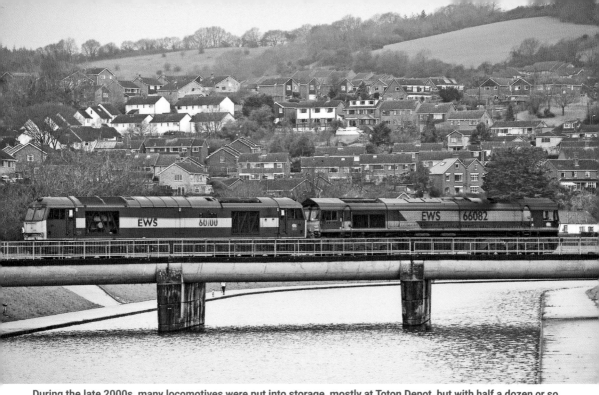

During the late 2000s, many locomotives were put into storage, mostly at Toton Depot, but with half a dozen or so initially finding their way down to St Blazey in deepest Cornwall of all places! However, when DB decided to embark on its 'Super 60' project most of these were resurrected. This is last built 60100 (formerly *Boar of Badenoch*) being towed by 66082 on the approach to Exeter St Davids as the 0Z48 St Blazey to Westbury, on the first leg of its journey to Toton. 8 March 2013.

This is 60100 just a year later than the previous image, and after the 'Super 60' treatment, powering through Tilehurst with 6B33, the 13.25 Theale to Robeston empty tanks. Although this is the last built loco, it was not the last in service, entering traffic on 9 December 1992 and named *Boar of Badenoch*, which it retained until May 2003. A month later, the locomotive was renamed *Pride of Acton*, which it then kept until March 2010 when it was stored. Returning to traffic, it ran nameless for a few years until June 2018 when it was named *Midland Railway - Butterley* during the line's diesel gala. It can sometimes be quite tricky keeping up with all these namings and denamings! 13 March 2014.

Crossing the River Usk at Newport, DB Schenker red 60044 *Dowlow* makes a fine sight with 6B41, the 11.15 Westerleigh to Robeston tanks. This loco was originally named *Ailsa Craig* when new. 24 February 2017.

Making a powerful view, this is 60071 *Ribblehead Viaduc*t powering up the gradient towards Cam & Dursley, just south of Gloucester, with 6E41, the 11.41 Westerleigh to Lindsey tanks. Originally entering service named *Dorothy Garrod*, it gained its most recent name from Class 47 47760, which went on to work for West Coast Railways. 60071 was put into storage in January 2016 and has not worked since. 5 March 2010.

Soon after passing through Hinton Admiral station, which is situated on the edge of the New Forest, this is 60079 *Foinaven* powering 6M57, the 14,00 Wool to Watford loaded sand train. This was one of the very last times EWS operated this train, it then passing to Freightliner, but it has not run for many years now. I can only assume there must be something drastically wrong with this particular loco, as it was refurbished as part of the 'Super 60' project but only worked for a while before being stored again in August 2015, and at the time of writing has not been in traffic since. 3 April 2008.

60053 *Nordic Terminal* is seen stabled at Didcot Parkway, while to the right just inside the gates of the Didcot Railway Centre is preserved main line-registered (at that time) Class 46 46035, which had just arrived with a parcels train from Bristol! 60053 entered traffic named *John Reith* and was put into storage back in August 2008; it is now slowly falling apart in Toton Yard. 14 May 2003.

Slowly passing through Didcot Parkway on the westbound local line, this is 60096 (originally *Ben Macdui*) with 6B33, the 13.35 Theale to Robeston tanks. The view from this location has now changed out of all proportion with the coming of the electrification and the construction of a huge multi-storey car park to the left of the picture. The loco was refurbished in 2014, firstly for use with Colas Rail but is now in the hands of GBRf and carrying the unofficial name of *Impetus*. 8 October 2009.

Another view of 60091 *Barry Needham*, this time approaching Southampton Airport Parkway station with 6B44, the 12.07 Southampton Western Docks to Eastleigh Yard. This short trip working often utilised a Class 60 if one was in the area. 18 September 2014.

A panoramic view of Bristol Temple Meads station with 60076 (originally *Suilven*) awaiting time with 6V54, the 06.14 Chirk Kronospan to Exeter Riverside empty log train. Unfortunately, this is yet another train that has since stopped running. The heaps of rubble in the foreground are all that is left of the former Bristol Bath Road diesel depot that closed back in 1995. The loco is now in operation with GBRf carrying the name *Dunbar*. 20 May 2015.

Transrail-liveried 60055 *Thomas Barnardo* is seen on the old stabling point at Godfrey Road, Newport, that has since been redeveloped, and views such as this are now impossible. Fortunately, the loco is still going strong having been refurbished for DCR and also retaining its original name. 24 September 2004.

Fast forward 16 years almost to the day and we find a now fully refurbished 60055 *Thomas Barnardo* passing Eastleigh station with 6Z91, the 13.56 Eastleigh Yard to Southampton spent ballast. It just shows what can be done to an apparently derelict machine with a little bit of time and money! 10 September 2020.

Seen here about a mile or so east of Bournemouth station, 60090 *Quinag* catches the low winter sunshine powering the heavily loaded 6M57, the 14.00 Wool to Watford loaded sand train. The area to the left of the picture was at one time Bournemouth East goods yard, which closed back in the 1960s near the end of steam operations. The loco was stored just two months after this image was taken in April 2008 and has not worked since. 12 February 2008.

In a strange, one-off mixture of Railfreight Coal and Transrail livery, 60066 (formerly *John Logie Baird*) is heading south through Millbrook with 7O48, the 09.23 Whatley Quarry to Hamworthy stone train. This was the last of the 'Super 60' overhauls for DB Schenker. 2 December 2003.

Another then and now loco comparison with the previous image. Almost exactly ten years on, and with one of its first workings since being refurbished, here is 60066 at Theale preparing to depart with 6B33, the 13.00 to Robeston. The silver livery is a one-off promoting Drax - Powering Tomorrow, although it does not actually carry a name. 17 December 2013.

Crossing a low viaduct along the causeway at Holes Bay, just west of Poole, is 60056 *William Beveridge* with 7V48, the 15.00 Hamworthy Quay to Whatley empty stone train. This is one of the last times a Class 60 worked this train, and after a more recent spell of running in the 2010s, it has since ceased once again. The loco was refurbished for Colas Rail in 2014, but has since moved to GBRf. 4 September 2008.

Now looking immaculate in Colas Rail livery, 60056 approaches Salisbury with 6V30, the 10.39 Eastleigh Yard to Westbury engineers' train. The loco has not received another name, official or unofficial, to date. 11 August 2015.

Passing the site of the former marshalling yard at Severn Tunnel Junction, refurbished 60063 (originally *James Murray*) is taking the Gloucester line with 6B13, the 05.00 Robeston to Westerleigh tanks. It is now difficult to visualise the amount of locos and freight that used to stable in the area immediately above the train. 20 April 2012.

This is 60052 *Glofa Twr - The Last Deep Mine In Wales - Tower Colliery* approaching the short tunnel under Holdenhurst Road on the approach to Bournemouth station with 7O48, the 09.23 Whatley Quarry to Hamworthy Quay stone train. The area where the coaches are parked to the right was once the site of the former Bournemouth East terminus station opened in 1870 when the line first reached Bournemouth, but closed in 1885 when the present through station opened. Surprisingly, the old station lingered on right up until the mid-1980s, finding further use as a warehouse and offices. Originally named *Goat Fell*, the loco was put into store in October 2008 and has not worked since. 9 March 2005.

In the low sunshine on a fabulous winter morning, 60091 crosses the River Usk bridge at Newport with 6B13, the 05.00 Robeston to Westerleigh tanks. 3 February 2012.

Having just passed Southcote Junction and approaching Reading West, 60059 *Swinden Dalesman* was in charge of the usual motley selection of wagons that formed 6M20, the 10.20 Whatley to St Pancras stone train, around this time. This view is now hindered by the overhead wiring. The loco has since undergone refurbishment and is still in everyday use with DB Cargo, retaining its nameplates. 21 January 2009.

Four years later and the transformed 60059 *Swinden Dalesman* is now in the DB Schenker red house colours as it passes the busy section of line between Southampton Maritime Freightliner Terminal and the Millbrook Freightliner Terminal with 6O41, the 10.14 Westbury to Eastleigh. The footbridge from which this image was taken has since been closed, deemed by the council to be unsafe. 1 May 2013.

The most recent operator to have acquired some of the class after undergoing the 'Super 60' treatment is Devon & Cornwall Railways (DCR). This is 60046 *William Wilberforce* passing Twyford with Chiltern DVT 82309 as 5Z20, the 08.14 Bristol Barton Hill to Wembley depot. I am pretty sure this is not the sort of train that the designers were envisaging when the locos were first conceived! It is rather nice that DCR has restored the original name to the loco. 1 February 2020.

Almost a picture postcard location, this is Colas Rail-operated 60076 (formerly *Suilven*) passing over the bridge at Greenland Mill on the approach to Bradford on Avon with the diverted 6V62, the 11.20 Tilbury to Llanwern empty steel train. This loco has since been taken on by GBRf. 18 April 2015.

This is 60017 *Shotton Works Centenary Year 1996* passing through Salisbury station with 6O41, the 10.14 Westbury to Eastleigh East Yard engineers' train. Owing to the many problems first encountered when bringing the class into service, this particular loco turned out to be the first accepted by BR on 30 October 1990. Originally named *Arenig Fawr*, it is still in service today with DB Cargo but does not carry a name at present. 18 January 2006.

60082 (formerly *Mam Tor*) takes the curve through Newport station with a westbound steel train heading for Llanwern. The loco was put into store in November 2008 but, unlike the rest of the withdrawn locos rusting away at Toton, it is currently doing the same thing at Crewe. 5 June 2003.

With smoke and fumes seemingly coming from everywhere, this is 60065 *Spirit of Jaguar* passing the South West Trains (now South Western Railway) depot at Northam with 6X44, the 12.07 Southampton Western Docks to Eastleigh East Yard trip working. Originally named *Kinder Low*, this unrefurbished loco is a bit of a survivor, somehow managing to stay in the active pool with DB, mainly for use as a yard shunter, although at the time of writing it is out of use. 21 August 2013.

With the winter sunshine already beginning to set at three in the afternoon, this is 60040 *The Territorial Army Centenary* in the attractive one-off maroon livery that it carried for a while, seen approaching Swindon station with 6B33, the 13.35 Theale to Robeston tanks. Originally named *Brecon Beacons*, the loco continues to do its duty with DB Cargo. 3 December 2008.

In original Railfreight Construction livery, 60099 *Ben More Assynt* pauses briefly for a signal check at Reading with an Acton to Merehead empty stone train. 1 March 1995.

Fast forward again almost 16 years later from the previous image and this is 60099 in silver Tata Steel livery (now un-named), seen waiting to depart the Westerleigh branch just south of Yate with 6E41, the 11.41 Westerleigh-Lindsey tanks. Unfortunately, the loco was stored in December 2015 and has not worked since. 18 February 2011.

Another shot of 60099, taken soon after the last image, passing the old GWR goods shed that formed part of the original Yate station as it gets underway northbound through the now rationalised and reopened (in 1989) Yate station with 6E41, the 11.41 Westerleigh to Lindsey tanks. 18 February 2011.

A more unusual duty for 60099 this time as it approaches Millbrook with 6B44, the 12.07 Southampton Western Docks to Eastleigh East Yard trip working. This is the aforementioned filling in turn at the time for locos stabling in Eastleigh Yard, with a Class 60 occasionally being used. 5 February 2013.

Before the station was totally rebuilt in the mid-2010s, this is Railfreight Construction-liveried 60039 *Glastonbury Tor* passing through Reading with an eastbound oil train, which in the days of sectorisation was unusual. It was probably heading for the Langley oil terminal near Slough. Thankfully the loco is still with us 30 years later and in everyday use with DB Cargo in its red livery, but carrying the rather less inspiring name of *Dove Holes*. 26 July 1991.

Even at this late date, still carrying the long defunct Railfreight Petroleum livery is 60054 *Charles Babbage* heading north out of Poole with 6M57, the 14.00 Wool to Watford sand train. More recently, the conveyance of sand from Wool went over to Freightliner haulage, but this has also now ceased. The loco is still in operation today with DB Cargo but not carrying a name. 20 March 2008.

Six years later and 60054 has now been refurbished and repainted in DB red livery. It is seen passing Bevois Park (now reopened as Southampton Up Yard), just south of St Denys, with 6Y32, the 08.24 Fawley to Holybourne fuel tanks. At the time it was unusual to see a Class 60 on this working, which stopped running in 2016 after many years of operation. In 2021, this loco has been modified to run on hydro-treated vegetable oil (HVO) fuel. 5 November 2014.

60010 (formerly *Plynlimon Pumlumon*) comes up the long straight section of line approaching Holton Heath, just east of Wareham, with 6M57, the 14.00 Wool to Watford loaded sand train. When new, this was the only loco to carry a two line nameplate; it is now un-named but has been refurbished and is still in service with DB Cargo. 5 February 2008.

When EWS took over freight operations in the UK in 1996, some locos were repainted into the maroon and gold house livery quite soon, but rather than undertake the costly exercise of mass painting all of its acquired locos, the company simply placed large stickers over the original sub-sector decals. Sporting this look, 60063 *James Murray* is seen exiting the passing loop at Haresfield, just south of Gloucester, with 6B13, the 05.00 Robeston to Westerleigh tanks, a reliably solid Class 60 working. The loco has since been refurbished by DB. 3 March 2011.

With the sun just starting to rise on a cold winter morning, this is a very rare working for one of the class as 60054 curves into Bristol Temple Meads station with 6V51, the 03.05 Warrington Arpley to Portbury Automotive Terminal empty car carriers. It is thought that this was the only available loco in the Warrington area at the time. So far, up to mid-2021, this area has escaped the electrification that was due to have been completed here in 2016/17, being curtailed because of spiralling costs. 18 December 2012.

60096 (originally *Ben Macdui*) is seen just south of Micheldever on the South Western Main Line with 6X01, the 10.18 Scunthorpe Yard to Eastleigh East Yard long-welded rail train that has since transferred to GBRf operation. Since the date of this image, the loco has also undergone a couple of changes, firstly being 'Super 60' overhauled for Colas Rail but then transferred to GBRf. 23 April 2010.

Five years later and a now refurbished 60096 is operating for Colas Rail and is seen here negotiating its way through Taunton with the empty wagons that form 6V54, the 05.35 Chirk Kronospan to Exeter Riverside empty log train. Yet another train that no longer runs, unfortunately. 60096 is now part of the GBRf fleet. 14 October 2015.

Carrying a one-off Cappagh blue livery, this one certainly stands out! Fully refurbished and now operated by DCR, 60028 (originally *John Flamsteed*) is recorded passing Twyford with a 0Z53 Bristol Temple Meads to Willesden route learning trip, which had run via Newbury on the Berks and Hants line. 7 April 2021.

A clean looking 60014 *Alexander Fleming* awaits its next duty at Westbury. As well as the addition of the large EWS sticker, it still retains the Railfreight Petroleum decals behind the cab door. This is one of the class officially withdrawn in recent times, but was actually stored back in December 2008. 1 March 2005.

This is 60015 (originally *Bow Fell*) approaching Swaythling with another unusual duty for a Class 60, 6V38, the 11.39 Marchwood to Didcot Yard MoD train. Despite this loco having been constructed during 1989, it was not until the end of March 1993 that it was finally accepted into traffic, becoming the last of the class to do so. 13 November 2012.

Differing styles of Colas Rail logos are seen on 60095 and 60076 stabled at Eastleigh. Both locos are now working for GB Railfreight with 60095 in full GBRf livery. 2 October 2015.

60052 *Glofa Twr – The Last Deep Mine In Wales – Tower Colliery,* with an interesting load of army vehicles, is on the approach to Eastleigh with 6V38, the 11.39 Marchwood to Didcot Yard. 60052 (originally *Goat Fell*) was stored just six months after this image was taken in October 2008. 3 April 2008.

With the EWS sticker having taken a bit of a battering, 60015 *Bow Fell* is seen again on a cold day with frequent snow showers, this time passing the well-known Battledown flyover. This is the junction of the Bournemouth and Exeter lines south of Basingstoke with the working being the 6O26 10.19 Hinksey Sidings to Eastleigh East Yard engineers' train. The loco is still hard at work with DB Cargo. 17 December 2010.

Three shots now showing the different guises of 60062 *Samuel Johnson*. Firstly, wearing its correct Railfreight Petroleum livery, this image shows the loco crossing from the down main line to the up main line at Wareham station with the 6V13 13.20 Furzebrook to Hallen Marsh LPG tanks. A well-known train for many years, this has long since stopped with the dedicated wagons being disposed of soon after. 11 August 1992.

Some 12 years later, and in immaculate ex-works EWS livery, 60062 is now un-named. It is seen awaiting its next turn of duty at Westbury. 8 February 2004.

Finally, another 14 years later and the same loco is now named *Stainless Pioneer* and wearing DB red livery. It is seen at Wyke Champflower, just east of Castle Cary, with 6C28, the 13.44 Exeter Riverside to Whatley Quarry stone empties. This train actually still runs but has recently moved over to Freightliner operation. 26 September 2018.

Seen from above the western portal of Southampton Tunnel, refurbished 60092 (originally *Reginald Munns*) passes through Southampton Central station with 6X44, the 12.07 Southampton Western Docks to Eastleigh East Yard trip working. At the time of writing, this view has now been blocked by vegetation. 23 April 2013.

With a much larger than usual three digit number on the front, a very work-weary looking 60068 *Charles Darwin* passes Westbury station with 6M20, the 10.20 Whatley Quarry to St Pancras stone train. This loco only lasted another eight months in traffic after this image was taken, being put into store during September 2009. 28 January 2009.

60072 *Cairn Toul* heads north under the distinctive varying height series of arches on the approach to Winchester with 6V38, the 11.39 Marchwood to Didcot Yard MoD train, yet another train long since lost. 60072 itself is just about lost as it was stored at Toton in February 2009 and has not worked again. 27 March 2003.

Taken from the nearby multi-storey car park, and with the still vast complex of Eastleigh Works in the background, 60010 (formerly *Plynlimon Pumlumon*) curves into Eastleigh station from the Portsmouth line with 7V16, the 11.47 Fareham to Westbury empty stone train, which to this day is still a runner but now with Freightliner motive power. 60010 is however still in use with DB Cargo. 4 November 2010.

Approaching Reading West, this is DB Schenker red 60017 (originally *Arenig Fawr*) with 6B33, the 13.35 Theale to Robeston tanks. This is still an almost solid Class 60 turn today, although a few years ago pairs of Class 66s were trialled without a great deal of success. 8 January 2013.

Still carrying EWS livery and unrefurbished, 60045 *The Permanent Way Institution* comes off the docks line and weaves its way through Southampton Central with 6B44, the 12.07 Southampton Western Docks to Eastleigh East Yard trip working. This particular name was once carried by Class 47 47644 while the name originally carried by 60045 was *Josephine Butler*. It was put into store in December 2014 after being used on and off as a shunting loco in various DB-operated yards. 3 September 2013.

This is 60024 (originally *Elizabeth Fry*) powering east through Twyford with 6M20, the 10.20 Whatley Quarry to St Pancras stone train. The loco has since been refurbished for DB Schenker, receiving its red livery and renamed *Clitheroe Castle*. 30 January 2008.

At almost the same spot as the previous image, 60001 (formerly *Steadfast*) is seen with the same train six years later. This loco also carried the name *The Railway Observer* between 2001 until its 'Super 60' refurbishment in 2014. 28 October 2014.

A nice clean 60009 (originally *Carnedd Daffyd*) heads south from Eastleigh with a short sand train bound for Wool, Dorset. This loco was stored in November 2010 and has been slowly rusting away ever since. Notice the Southern Region slam-door electric unit, which were in the process of being rapidly withdrawn at the time, approaching the station. 27 August 2003.

60045 *The Permanent Way Institution* is seen again, this time under a rather dark sky pulling away from a crew change at Didcot Parkway with a Hayes & Harlington to Moreton on Lugg stone train. Notice the Immingham 40B steam loco style shedplate sticker above the coupling hook. 2 November 2012.

Looking very smart in EWS maroon livery and with a unique style of nameplate, 60089 *The Railway Horse* passes light engine through Newport station. Originally named *Arcuil* when new, the loco was consigned to store at Toton in 2008 and has not worked since. 5 June 2003.

With St John's Church of England prominent in the background, 60092 (formerly *Reginald Munns*) heads west at Lockerley (between Romsey and Salisbury) with 7V16, the 11.47 Fareham to Westbury empty stone train. This service is now in the hands of Freightliner locomotives. 25 April 2013.

Another unusual working for one of the class. 60024 (formerly *Elizabeth Fry*) is seen at Andover, shunting a cable-laying train from Eastleigh Yard to MoD Ludgershall. From Andover station, the couple of miles or so to Ludgershall (to serve the MoD depot) is now the current limit of the once cross-country Midland & South Western Junction Railway route from Cheltenham to Southampton via Swindon, which would be rather handy these days but is now long since closed. 19 October 2012.

Another view of 60045 *The Permanent Way Institution*, this time just east of Farnborough (Main) on the South Western Main Line with 6O12, the 03.23 Merehead to Woking stone train. At this time, Class 60s were very scarce on this section of line, and today they are just about non-existent. 13 September 2013.

The Permanent Way Institution

60045

Complete with Mainline branding, a very clean 60042 *Dunkery Beacon* heads east through Kensington Olympia with an ARC stone train bound for Acton. 4 June 1996.

Eight years later than the previous image and 60042, now named *The Hundred of Hoo*, passes through Cardiff Central with an empty steel train heading for Margam. South Wales was always a good place to see these locos at work, and to a somewhat lesser extent is one of the few areas that still is today. Unfortunately, the loco was stored in November 2008 and has not worked since. 24 September 2004.

60053 *Nordic Terminal* comes around the sharp curve at Salisbury with 7V16, the 11.47 Fareham to Westbury empty stone train. Entering traffic in 1991, the loco was originally named *John Reith* and, like so many others around the time, it was put into store during August 2008. 14 February 2006.

60013 *Robert Boyle* has just departed Westbury and is approaching Lambert's Bridge with an unidentified westbound empty stone train for Merehead Quarry. The class was getting pretty scarce in the Westbury area around this time, and in fact this loco was withdrawn just two months after the date of this image in May 2011. 3 March 2011.

60056, 70801, 70805, 70807, 70811 and 70816 are pictured during a shunting move at Westbury. When operating with Colas Rail, examples of the company's Class 60 fleet were not that common at Westbury, despite there being a servicing point located there. 15 August 2017.

Seen from high above the Great Western Main Line, this is 60096 (originally *Ben Macdui*) near Cholsey with 6B33, the 13.35 Theale to Robeston tanks. Overhead electrification has now ruined this once fine view. 60096 is now at work for GBRf. 17 March 2011.

During 2018, all the former Colas Rail locos were taken on by GBRf. This is 60095 (originally *Crib Goch*) passing Woking on its final trip In Colas Rail colours with one wagon forming 6Z95, the 09.00 Whitemoor Yard to Eastleigh Works. It was released from the works a few months later in full GBRf livery. 31 August 2018.

Absolutely immaculate 60026 *Helvellyn* takes a short test trip within the confines of the Eastleigh Works complex just after being repainted in an attractive blue livery for GBRf. Interestingly, the new nameplates are in the style of the original Class 44 Peak locomotives, this particular one is a replica of that originally attached to 44002. It is a great touch from GBRf but might have been even better had the company attached them to 60002 instead, which is another GBRf loco. 2 October 2019.